I0102014

AI, ALIENS & CONSPIRACIES
THE TRUTHFUL ANALYSIS

Jeremy Griffith

www.HumanCondition.com

AI, Aliens & Conspiracies: the truthful analysis by Jeremy Griffith

Published in 2023, by WTM Publishing and Communications Pty Ltd
(ACN 103 136 778) (www.wtmpublishing.com).

All enquiries to:

WORLD TRANSFORMATION MOVEMENT®
Email: info@worldtransformation.com
Website: www.humancondition.com or www.worldtransformation.com

The World Transformation Movement (WTM) is a global not-for-profit movement
represented by WTM charities and centres around the world.

ISBN 978-1-74129-094-3
CIP – Biology, Philosophy, Psychology, Health

Commendations for Griffith's treatise

From Thought Leaders

'[**Prof. Stephen Hawking**] is most interested in your impressive proposal.'
● 'In all of written history there are only 2 or 3 people who've been able to think on this scale about the human condition.' **Dr Anthony Barnett**, Prof. of Zoology
● '*FREEDOM* is the book that saves the world…cometh the hour, cometh the man.' **Prof. Harry Prosen**, Pres. Canadian Psychiatric Assn. ● 'I am stunned and honored to have lived to see the coming of "Darwin II".' **Prof. Stuart Hurlbert**, esteemed ecologist ● 'Living without this understanding is like living back in the stone age, that's how massive the change it brings is!' **Prof. Karen Riley**, clinical pharmacist ● 'Frankly, I am blown away by the ground-breaking significance of this work.' **Dr Patricia Glazebrook**, Prof. of Philosophy ● 'I've no doubt a fascinating television series could be made based upon this.' **Sir David Attenborough** ● '*FREEDOM* is the necessary breakthrough in the critical issue of needing to understand ourselves.' **Dr David J. Chivers**, former Pres. Primate Society of Britain ● 'Whack! Wham! I was converted by Griffith's erudite explanation for our behaviour.' **Macushla O'Loan**, *Executive Women's Report* ● 'This is indeed impressive.' **Dr Roger Lewin**, preeminent science writer ● 'I have recommended Griffith's work for his razor-sharp biological clarifications.' **Dr Scott Churchill**, Prof. of Psychology ● 'An original and inspiring understanding of us.' **Dr Charles Birch**, Prof. of Zoology ● 'The insights are fascinating and pertinent and must be disseminated.' **Dr George Schaller**, preeminent biologist ● 'Very impressive, particularly liked the primatology section.' **Dr Stephen Oppenheimer**, geneticist, author *Out of Eden* ● 'I consider the book to be the work of a prophet.' **Dr Ron Strahan**, former dir. Sydney Taronga Zoo ● 'The scholarly value [of Griffith's synthesis] is comparable to several of the most celebrated publications in biology.' **Prof. Walter Hartwig**, anthropologist ● 'I believe you're on to getting answers to much that has bewildered humans.' **Dr Ian Player**, famous Sth. Afr. conservationist ● 'A superb book, a forward view of a world of humans no longer in naked competition.' **Dr John Morton**, Prof. of Zoology ● 'This might bring about a paradigm shift in the self-image of humanity.' **Dr Mihaly Csikszentmihalyi**, Prof. of Psychology ● 'As a therapist this is a simply brilliant explanation.' **Jayson Firmager**, founder of *Holistic Therapist Magazine* ● 'The questions you raise stagger me into silence; most admirable.' **Ian Frazier**, author *Great Plains* bestseller ● 'The WTM is an island of sanity in a sea of madness.' **Tim Macartney-Snape**, world-leading mountaineer & twice Order of Australia recipient

Commendations From The General Public

'Griffith should be given Nobel prizes for peace, biology, medicine; actually every Nobel prize there is!' ● 'He nailed it, nailed the whole thing, just like the world going from FLAT to ROUND, BOOM the WHOLE WORLD CHANGES, no joke.' ● '*FREEDOM* will be the most influential, world-changing book in history, and time will now be delineated as BG, before Griffith, or AG, after Griffith.' ● 'I'm speechless – this is bigger than natural selection & the theory of relativity!' ● 'I really think this man will become recognized as the best thinker this world's ever seen, and don't we need him right now!' ● 'Griffith has decoded the human species, we FINALLY know what's going on & the suffering stops!' ● 'The world can't deny this for much longer, let the light in, save the human race!' ● 'This is the most exciting moment in my life. *THE Interview* tore my hat off & let my brain fly into the sky!' ● '*THE Interview* should be globally broadcast daily. The healing explanation humans so sorely need.' ● 'In a world that's lost its way there's no greater breakthrough, water to a world dying of thirst.' ● 'Dawn has come at Midnight! A brilliant exposition, we could be on the cusp of regaining Paradise!' ● 'This man has broken the great silence, defeated our denial, got the truth up, woken us from a great trance.' ● 'Beware the 'deaf effect; your mind will initially resist the issue of our corrupted condition and so find it hard to take in or hear what's being said, but if you're patient you'll find the redeeming explanation of our condition pure relief.' ● 'John Lennon pleaded "just give me some truth", well this site finally gives us *all* the truth!' ● '*FREEDOM* is the most profound book since the Bible, now with the redeeming truth about us humans.' ● '*Death by Dogma* is brilliant clarification.' ● 'We were given a computer brain, but no program for it; but Aha, Griffith has found it, made sense of our lives!' ● 'This just goes deeper & deeper in explaining us, like dawn devouring darkness, amazing!' ● 'Agree, this is not another deluded, pseudo idealistic, PC, 'woke', false start to a better world, but the human-condition-resolved real solution.' ● 'Freedom indeed! What we have here is the second coming of innocence who exposes us but sets us free!' ● 'As prophesised, King Arthur has returned to save us (mentioned in par.1036 *Freedom*)' ● 'We all need to go back to school & learn this truthful explanation of life.' ● 'Join in our jubilation, your magic reunites, all men become brothers, all good all bad, be embraced millions! This kiss [of understanding] for the whole world' – From Beethoven's 9th (par.1049 *Freedom*)

Contents

Background

Jeremy Griffith is an Australian biologist who has dedicated his life to bringing redeeming and psychologically healing biological understanding to the dilemma of the human condition—which is the underlying issue in all human life of our species' extraordinary capacity for what has been called 'good' and 'evil'.

Jeremy has published over ten books on the human condition, including:

— *Beyond The Human Condition* (1991), his widely acclaimed second book;

— *A Species In Denial* (2003), an Australasian bestseller;

— *FREEDOM: The End Of The Human Condition* (2016), his definitive treatise;

— *THE Interview* (2020), the transcript of acclaimed British actor and broadcaster Craig Conway's world-changing and world-saving interview with Jeremy about his book *FREEDOM*;

— *Death by Dogma: The biological reason why the Left is leading us to extinction, and the solution* (2021), which presents the biological reason why Critical Theory threatens to destroy the human race;

— *The Great Guilt that causes the Deaf Effect* (2022), which describes how lifting the great burden of guilt from the human race initially causes a 'Deaf Effect' difficulty taking in or 'hearing' what's being presented;

— *The Shock Of Change that understanding the human condition brings* (2022), which addresses how to manage the shock of change that inevitably occurs when the redeeming understanding of our corrupted condition arrives;

— *Therapy For The Human Condition* (2023), which is about the therapy that is desperately needed to rehabilitate the human race from our psychologically upset state or condition, elaborating on what is presented in *FREEDOM*;

— *Our Meaning* (2023), which explains how being able to know and fulfil the great objective and meaning of human existence finally ends human suffering; and

— *The Great Transformation – How Understanding The Human Condition Actually Transforms The Human Race* (2023), which gives a concise description of how the psychological rehabilitation of humans occurs, and how everyone's life can immediately be transformed.

<u>This book, ***AI, Aliens & Conspiracies: the truthful analysis***, is a human-condition-confronting-not-avoiding , truthful analysis of the dangers of Artificial Intelligence and the possibility of Aliens visiting Earth.</u>

Jeremy's work has attracted the support of such eminent scientists as the former President of the Canadian Psychiatric Association Professor Harry Prosen, the esteemed ecologist Professor Stuart Hurlbert, Australia's Templeton Prize-winning biologist Professor Charles Birch, the Former President of the Primate Society of Great Britain Dr David Chivers, Nobel Prize-winning physicist Stephen Hawking, as well as other distinguished thinkers such as the pre-eminent philosopher Sir Laurens van der Post.

Jeremy is the founder and a patron of the World Transformation Movement (WTM)—see www.HumanCondition.com.

AI, ALIENS & CONSPIRACIES:
the truthful analysis

Introduction

[1] As I explain in my main book *FREEDOM*, and briefly summarise in *THE Interview*, human behaviour has been horrifically affected by the guilt and shame of us humans having corrupted our original cooperative, selfless and loving instinctive self or soul after we became conscious some 2 million years ago and became competitively, selfishly and aggressively behaved. As the American writer and philosopher Ralph Waldo Emerson described our dreadful situation, **'man is a god in ruins'**, **'insane and rabid'**; **'fallen men'** who **'plead…to return to paradise'** *(Nature*, 1844). This seeming original crime or 'sin' of becoming corrupted has made us seek reinforcement for ourselves from every situation we encounter, and attack and block out any criticism of ourselves. We have been *immensely* insecure egocentric seekers of power, fame, fortune and glory, and *extremely* aggressive and *deeply* alienated sufferers of the psychologically upset state of *the human condition*!

[2] As is presented in chapter 3 of *FREEDOM* and summarised in *THE Interview*, the redeeming 'instinct vs intellect' explanation of the human condition—which explains that when our nerve-based, understanding-dependent, self-adjusting, fully conscious mind emerged some 2 million years ago it had to defy our gene-based, naturally-selected instinctive orientations in order to fulfil its great potential to manage life—has now been found. However, until that explanation becomes widely known everything to do with us humans will remain *hugely* influenced by the effects of this extreme

psychological insecurity about our corrupted condition, and so it follows that our development of superintelligent forms of artificial intelligence or AI will *also* be hugely affected by that insecurity, and likely in extremely dangerous ways—and, as I will now explain, as a result of the extreme insecurity of our corrupted human condition AI *is* becoming extremely dangerous in a number of ways.

[3] In explaining this potential extreme danger from AI, I will also look at the possibility of an alien, extraterrestrial form of superintelligence visiting Earth and intervening in our lives. So this presentation, titled *AI, Aliens & Conspiracies: the truthful analysis*, will be like a discussion between us humans, AI and aliens—hence my cover drawing of an imagined discussion that takes place during a picnic attended by a human, an AI machine and an alien.

Since a superintelligent form of AI, or an alien visitor to Earth, would recognise the integrative meaning of existence, they would not be competitive, selfish or aggressive towards humans

[4] Recently *The Spectator* magazine (15 July 2023) featured an article about AI titled 'We need to be terrified for our lives' written by James W. Phillips (a former Special Adviser to the British Prime Minister for science and technology) and Eliezer Yudkowsky (head of research at the Machine Intelligence Research Institute in California), in which Yudkowsky worries that if we create 'a highly superintelligent form of [artificial] intelligence...they might calculate that they can get more of the galaxy for themselves if they stop humanity'. In his famous 1898 science fiction novel *War of the Worlds*, H.G. Wells actually describes a conflict between humans and presumably, if they're capable of travelling between worlds, highly intelligent extraterrestrial beings—and I might say that given the almost endless

number of other planets in the universe it would seem inevitable that life, and probably some fully conscious, highly intelligent life capable of vast interplanetary travel, would exist out there in the universe. <u>As I am going to explain, what is so extremely wrong about these human-condition-avoiding, Plato's-cave-hiding, dishonest anticipations of a conflict occurring between us and **'a highly superintelligent'** computer, or a highly advanced form of alien intelligence, is that such advanced forms of intelligence would recognise the very obvious truth of the integrative meaning of existence and as a result would not want to behave divisively and competitively, selfishly and aggressively try to **'get more of the galaxy for themselves'**, or **'war'** with our **'world'**.</u>

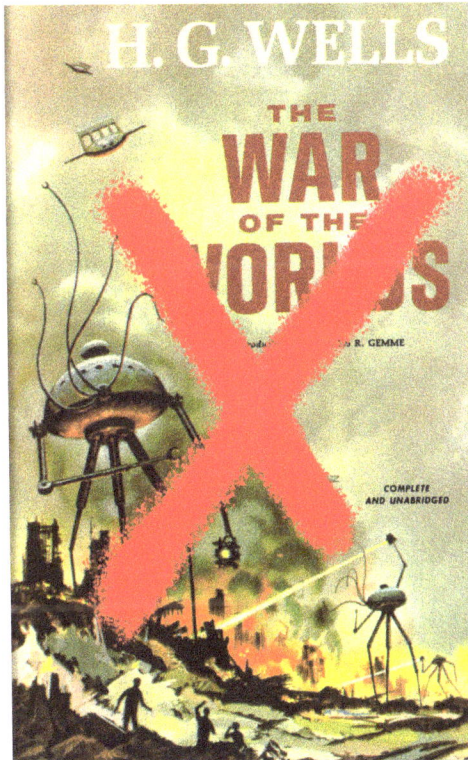

[5]In chapter 4 of *FREEDOM*, which is titled 'The Meaning of Life', and in its summary in Freedom Essay 23, I explain this obvious truth of the 'integrative meaning of existence' and how, until now, lacking the redeeming explanation of our corrupted, divisive, seemingly non-integrative, competitive, selfish and aggressive human condition, we have had no choice but to deny the seemingly condemning truth of Integrative Meaning. I begin that chapter and essay by pointing out that if you "take a look at our surroundings... you'll note that the most obvious characteristic of our world is that it is full of 'things', variously enduring arrangements of matter, like plants, animals, clouds and rocks. And not only that, it is apparent that these arrangements of matter consist of a hierarchy of ordered parts; a tree, for instance, is a hierarchy of ordered matter—it has a trunk, limbs, roots, leaves and wood cells. Our bodies are also a collection of parts, as are clouds and rocks, which are built from different elements and compounds. Furthermore, what we have seen happen over time to these arrangements of matter is that there has been a progression from simple to more complex arrangements. From the fundamental ingredients of our world of matter, space and time, matter has become ordered into ever larger (in space) and more stable or durable (in time) arrangements. [par. 314 of *FREEDOM*]

[6]To elaborate, our world is constructed from some 94 naturally occurring elements that have come together to form stable arrangements. For example, two hydrogen atoms with their single positive charges came together with one oxygen atom with its double negative charge to form the stable relationship known as water. Over time, larger molecules and compounds developed. Eventually macro compounds formed. These then integrated to form virus-like organisms, which in turn came together or integrated to form single-celled organisms that then integrated to form multicellular organisms, which in turn integrated to form societies of single species that continue to integrate to form stable, ordered arrangements of different species. Clearly, what is happening on Earth is that matter is integrating into

larger and more stable wholes. And this development of order is not only occurring here, it is also happening out in the universe where, over the eons, a chaotic cosmos continues to organise itself into stars, planets and galaxies. As two of the world's greatest physicists, Stephen Hawking and Albert Einstein, have said, respectively, **'The overwhelming impression is of order...**[in] **the universe'** (Gregory Benford, 'The time of his life', *The Sydney Morning Herald*, 27 Apr. 2002; see www.wtmsources.com/170), and **'behind everything is an order'** (*Einstein Revealed*, PBS, 1997). [par. 315]

[7]... The law of physics that accounts for this integration of matter is known as the 'Second Path of the Second Law of Thermodynamics', or 'Negative Entropy', which states that in an open system, where energy can come into the system from outside it (in Earth's case, from the sun, and, in the case of the universe, from the original 'big bang' explosion that created it), matter integrates; it develops order. Thus, subject to the influence of Negative Entropy, the 94 elements from which our world is built develop ever larger and more stable wholes. [par. 316]

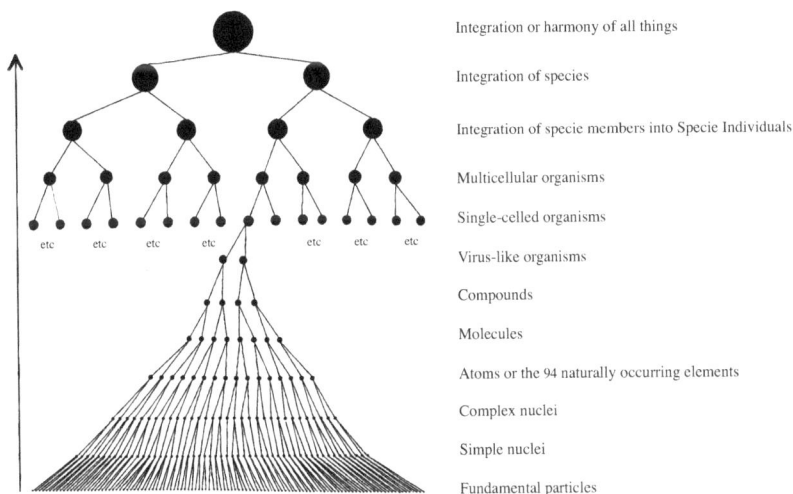

Chart showing the ordered integration of matter on Earth
(a similar chart appears in Arthur Koestler's book *Janus: A Summing Up*).

[8] ... <u>But while the integrative meaning of existence *is* the most</u> <u>obvious of all truths, it has also been the most difficult of all truths</u> <u>for humans to acknowledge, for an *extremely* good reason.</u> [par. 319]

[9] <u>The difficulty arises from the fact that for a collection of</u> <u>parts to form and hold together, for matter to integrate, the parts of</u> <u>the developing whole *must* cooperate, behave selflessly, place the</u> <u>maintenance of the whole above the maintenance of themselves,</u> <u>because if they don't cooperate—if they compete, behave selfishly or</u> <u>inconsiderately—then the whole disintegrates, the parts break down</u> <u>into the more elementary building blocks of matter from which they</u> <u>were assembled.</u> ... A leaf falling from a tree in autumn does so to ensure the tree survives through winter and carries on; it puts the maintenance of the whole, namely the tree, above the maintenance of itself. The effective functioning of our body similarly depends on the cooperation of all its parts, on every part doing what is best for the whole body. Our skin cells, for example, are in constant turnover, with new cells replacing the old ones that have sacrificed themselves to protect our body. [par. 320]

[10] ... <u>Put simply, selfishness is divisive or disintegrative while</u> <u>selflessness is integrative—it is the glue that holds wholes to-</u> <u>gether; it is, in fact, the theme of the integrative process, and thus</u> <u>of existence.</u> It is also what we mean by the word 'love', with the old Christian word for love being **'caritas'**, meaning charity or giving or selflessness (see Col. 3:14, 1 Cor. 13:1-13, 10:24 & John 15:13). So 'love' is cooperative selflessness—and not just selflessness but *unconditional selflessness*, the capacity, if called upon, to make a full, self-sacrificing commitment to the maintenance of the larger whole. <u>BUT—and herein lies the nub of the problem—if the meaning</u> <u>of existence is to behave *integratively*, which means behave co-</u> <u>operatively and selflessly, *why* do humans behave in the completely</u> <u>opposite way, in such a competitive and selfish *divisive* way? Yes,</u> <u>the integrative theme of existence squarely confronts us humans</u> <u>with the issue of the human condition, the issue of our non-ideal</u>

behaviour. And so despite being such an obvious truth, Integrative Meaning has been *so* horrifically condemning of the competitive, aggressive and selfish human race that until we could explain the *good reason why* humans have been divisively rather than integratively behaved (which was done [with the 'instinct vs intellect' explanation] in chapter 3)—and thus make it psychologically safe to admit the truth of the order developing, integrative meaning of existence—we had no choice but to live in near total denial of it. Hawking's, Einstein's…acknowledgments of the order developing, integrative process…were bold indeed. [par. 321]

[11] … [Lacking this redeeming explanation of our corrupted condition] the Negative Entropy-driven integrative, cooperative, loving, selfless, order-developing theme or meaning or purpose of existence has been an almost completely unconfrontable truth for the psychologically upset, competitive, aggressive and selfish human-condition-stricken human race. In fact, we have lived in *such* terrified fear and awe of the truth of Integrative Meaning, we have been *so* confronted, condemned and intimidated by it, *so* unable to deal with it on any sort of an equal footing, that we deified the concept—and not just as *a* God, but *the one and only* God, *the* most universal and fundamental, yet completely unconfrontable, of truths. [par. 324] [Thus, as I explain in par. 326, deifying Integrative Meaning as the religious concept of God made it a safely abstract, undefined concept that allowed us to avoid having to directly confront the truth of Integrative Meaning.]

[12] … [It is only] with the human condition now explained [by the 'instinct vs intellect' explanation presented in chapter 3 of *FREEDOM* and summarised in *THE Interview*] and our divisive, seemingly non-integrative state finally understood, [that] *all* humans can at last safely admit and recognise that there has only been one God, one all-dominating and all-pervading theme or meaning of existence, which is Integrative Meaning—a truth we recognise when we say **'God is love'** (Bible, 1 John 4:8, 16). [par. 327]"

[13] For a more complete description of Integrative Meaning read all of chapter 4 of *FREEDOM*.

[14] So we live in a Negative Entropy-driven integrative, cooperative, selfless and loving world. The two apparent exceptions to this are firstly our competitive, selfish and aggressive human condition, which, the 'instinct vs intellect' conflict in us finally explains the good reason for. As mentioned, I explain in chapter 3 of *FREEDOM* (and summarise in *THE Interview*) that when our nerve-based, understanding-dependent, self-adjusting, fully conscious mind emerged some 2 million years ago it had to defy our gene-based, naturally-selected, cooperative, selfless and loving instinctive orientations in order to fulfil its great potential to manage life. And this necessary defiance of our instincts by our conscious mind unavoidably caused us to become competitive, selfish and aggressive, but now that we have found understanding of why we became divisively behaved we can return to being integratively behaved. So our divisive behaviour has not *actually* been non-integrative, an exception to Integrative Meaning, which is why it was only an 'apparent' exception.

[15] The second apparent exception is the competitive, selfish and aggressive behaviour amongst non-human animals. This divisive behaviour among non-human animals occurs because the natural selection process normally can't develop unconditionally selfless behaviour—outside the nurturing, 'Love-Indoctrination' process that, as I explain in chapter 5 of *FREEDOM* and summarise in *THE Interview*, our bonobo-like ape ancestors were able to develop and by so doing create our cooperative, selfless and loving moral instincts. As I explain in chapter 4 of *FREEDOM*, "all non-human animal species are stuck in the 'animal condition', with each sexually reproducing individual member of the species forever having to compete to ensure its genes reproduce and carry on. *That* is the essential fact or rule of the gene-based natural selection process—genes are unavoidably selfish; they have to ensure they reproduce if they are to carry on. It is important to reiterate, however, that even though

this selfishness—and the extreme competition between the sexually reproducing individuals it gives rise to—is characteristic of virtually all of nature, such selfishness is *only* occurring because of the *limitation* of the genetic process of *normally* being unable to develop unconditional selflessness between sexually reproducing individuals. In his 1850 poem *In Memoriam*, [Lord Alfred] Tennyson famously wrote: **'Who trusted God was love indeed / And love Creation's final law / Tho' Nature, red in tooth and claw / With ravine** [in violent contradiction], **shriek'd against his creed.'** While Integrative Meaning or **'God'** and its theme of unconditional selflessness or **'love'** is the **'creed'** or **'final law'** of **'creation'** that the competitive, selfish and aggressive, **'red in tooth and claw'** characteristic of so much of **'Nature'** seems to be in violent contradiction **'against'**, we can now understand that this selfish characteristic doesn't mean that the *overall* biological reality of existence—life's meaning and theme—is to be selfish, as the dishonest [biological] theories of Social Darwinism, Sociobiology, Evolutionary Psychology and Multilevel Selection would have us believe. As will be explained in chapter 5 [of *FREEDOM* when the nurturing, Love-Indoctrination process that gave rise to our instinctive orientation to behaving unconditionally selflessly is explained], in the case of humans, we don't have selfish instincts like other species, rather we have *unconditionally selfless* instincts. And the selfishness that is characteristic of so much of nature is *only* occurring because of the *limitation* of the gene-based refinement process—its inability, in most situations, to develop unconditional selflessness. The genetic process would develop unconditionally selfless, fully cooperative behaviour between all sexually reproducing individuals if it could—because such selflessness is what is required to maintain a fully integrated whole—but, because of its particular *limitation*, it normally can't. Integrative selflessness, not divisive selfishness, is the *real* nature or characteristic of existence, the theme of life. [par. 357]" So the competitive, selfish and aggressive 'animal condition' doesn't actually contradict the truth of Integrative Meaning, it merely describes a difficulty that occurred in the gene-based development of the integration of matter. Therefore, like our divisive 'human condition', it only 'apparently' contradicts Integrative Meaning.

[16] As I also explain in *FREEDOM* and summarise in *THE Interview*, while denying the truth of Integrative Meaning and the nurtured origins of our cooperative instincts, and lacking the 'instinct vs intellect' explanation for why we corrupted that cooperative state, we misused the fact that non-human animals are competitive, selfish and aggressive to justify our divisive competitive, selfish and aggressive behaviour. As I summarise in paragraph 64 of *THE Interview*: "false as it [the idea that we have must-reproduce-our-genes, savage, competitive, selfish and aggressive instincts like other animals] is, it's been an absolutely brilliant excuse because instead of our instincts being all-loving and thus unbearably condemning of our present non-loving state, they are made out to be vicious and brutal must-reproduce-your-genes instincts like other animals have; *and*, instead of our conscious mind being the instinct-defying cause of our corruption, it was made out to be the blameless mediating 'hero' that had to step in and try to control those supposed vicious instincts within us!"

[17] What this all means is that we humans have been living in denial of Integrative Meaning and dishonestly believing that selfish competition and aggression is the characteristic of existence, hence the very wrong assumption that a superintelligent form of AI will want to competitively, selfishly and aggressively **'get more of the galaxy for themselves'**, or even competitively, selfishly and aggressively **'want to overpower and enslave us'** as many other reviewers of AI have suggested—or that a highly advanced form of alien intelligence will want to competitively, selfishly and aggressively **'war'** with our **'world'**! As I said, and have now explained, such thinking is extremely false, human-condition-denying, Plato's-cave-hiding thinking. To avoid the truth of our corrupted human condition we denied the truth of Integrative Meaning and dishonestly maintained that the competitive, selfish and aggressive behaviour we humans practice is

a universal practice and therefore that a superintelligent machine or aliens who are intelligent enough to be able to visit us would think competitively, selfishly and aggressively when they would actually recognise the obvious truth of the integrative meaning of existence and be appreciative of the need to be integrative, cooperative, selfless and loving.

Would AI or superintelligent aliens intervene in our world?

[18] A superintelligent form of artificial intelligence, or superintelligent beings elsewhere in the universe, are not going to want to competitively eliminate or dominate or enslave humans, they are going to want to be cooperative, selfless and loving. So they will try to work out how to integratively cooperate with us. And working out how best to integratively get along with us, the superintelligent AI or alien will have to take into consideration the fact that we are a part of this planet, like all the other animals, plants and organisms here are, while they are a foreign, new presence on Earth. The integrative question then for them, and us, to think about is would they intervene in our lives and world, or leave us and our world alone?

[19] This is actually the same question we humans will eventually face if and when we heal our psychosis with the understanding we now have of our corrupted condition and become cooperative, selfless and loving again: will we want to intervene in the development of the lives of other creatures on Earth? For example, will we selectively or directly alter their genetic make-up so they can overcome their ferociously competitive, have-to-reproduce-their-genes 'animal condition' (that I explained earlier) and even become fully conscious like we are—or will we leave them to continue living the way they are? I think we would decide not to intervene in their development—apart from doing such things as trying to stop rogue

plants (like the Melaleuca invading the Everglades in Florida) or animals (like the cane toad invasion of Australia) or microorganisms (like deadly viruses) that cause terrible destruction on Earth. For one thing, since our original instinctive self or soul grew up with the other forms of life here the way they are, we need the presence of how they are in our lives for our psychological well-being.

[20] For the same sort of reasons, I don't think selfless superintelligent forms of AI or aliens would want to intervene and get involved in our development. Certainly not in a divisive, non-integrative way because they would appreciate the integrative meaning of existence, but not even in a helpful, cooperative, get-along-with-us integrative way because, as I say, I think they would not want to intervene in our development. Clearly we humans are going through a stage where we are extremely psychotic, but even if we were to ask these forms of superintelligence to help us with our psychosis-stricken condition, outside of maybe agreeing to help us avoid doing something totally self-destructive, just as we try to stop destruction from out-of-control rogue organisms, I think they might well not want to interfere in our development and refuse to help us, and instead leave us to find our own way forward, just as we might well decide not to intervene and stop other animals having to endure the 'animal condition'. This, by the way, means I don't think we will find any evidence of alien extraterrestrial superintelligent beings visiting us even if such intelligent beings exist out there in the cosmos and have visited us (because, again, I don't think they would want to intervene in and intrude into our lives or world).

[21] This question of what we might do when we become free of psychosis (such as whether we should free other creatures on Earth from their ferociously competitive 'animal condition', or assist them to become fully conscious) is obviously best left to that future psychosis-free, sound-thinking situation, but this still leaves us having to try to work out what superintelligent AI might do in the immediate future—whether it might want to dramatically intervene in some way in human progress, or in the development

of any of life on Earth. As I will conclude at the end of this essay, I think there are so many dangers with AI that we shouldn't develop it further until we have become a psychosis-free species, and if we do halt its development it will mean that we won't at the present time have to address this question about whether AI might want to dramatically intervene in our lives or in life on Earth.

Our great fear of AI recognising the truth of Integrative Meaning and exposing our corrupted condition

[22]The next point to make now that we have at last explained the human condition, explained the very good reason we've had to be divisively behaved, and as a result are able to admit the truth of Integrative Meaning, is to recognise the existence of a previously hidden deep fear we humans have of AI—which I strongly suspect an early report about computers was alluding to when it said, **'Mankind has long been…frightened by the prospect of creating machines that think'** (*Newsweek*, 4 Jul. 1983). This deep fear is of AI becoming intelligent enough to realise the obvious truth of Integrative Meaning and as a result condemn humans for being divisively, competitively, selfishly and aggressively behaved. Although humans have been living in denial of Integrative Meaning, it is *such* an obvious truth we are all actually subconsciously aware of it, and, as a result, subconsciously *deeply* afraid of its implied condemnation of us. After all, this subconscious awareness and resulting fear of Integrative Meaning is why we attempted to make Integrative Meaning something totally unrelated to us by deifying it as 'God'. So yes, while we haven't been able to admit and talk about it, one of the main reasons for the extreme fear, indeed paranoia, we humans have had about AI is that it will reveal the truth of Integrative Meaning and thus condemn us as being guilty, bad, evil, sinful monsters; unbearably accentuate the agony of our corrupted human condition! Clearly, finding the

redeeming, good reason for our corrupted condition will bring an end to this underlying fear of Integrative Meaning and finally relieve us of our fear of AI exposing us as being evil, worthless, out-of-step-with-creation beings—thus allowing us to think about AI in a far more balanced, paranoia-free, truthful way.

[23] In the Addendum at the end of this essay I will talk more about this subconscious fear we humans have had of the truth of our corrupted condition and how our resulting determined denial of that truth left us sensing that there is some great truth we can't access, leading us to come up with all manner of mad conspiracy theories to try to explain that hidden truth. In that Addendum I'll also explain why we have come to portray aliens as having the neotenous features of large eyes, snub noses and domed foreheads.

The danger of AI having a left-wing bias, and the danger of a 'half smart' computer not understanding the paradox of the human condition

[24] The next issue about AI that arises from our historic insecurity about our corrupted human condition has to do with left-wing bias. Having lived in denial of our corrupted condition has meant we haven't acknowledged that the ultimate purpose and objective of human progress has been to find understanding of it—clearly, if there is no corrupted condition then there is no corrupted condition to have to be explained! And, as I will now describe, not acknowledging this purpose of having to find understanding of our corrupted condition has made the ideology of the left-wing in politics of dogmatically imposing cooperative, selfless and loving behaviour virtually impossible to argue against—which, as we will see, has allowed advocates of left-wing ideology to dangerously misuse technology such as AI.

[25] As I explain in my booklet *Our Meaning*, in Aspect 3 that is titled 'Ending the polarised world of politics', "while there has been no admission of our purpose and meaning of having to carry out the immensely psychologically upsetting, heroic search for knowledge, ultimately self-knowledge, understanding of our corrupted human condition, there has been no justification for the upset, competitive, selfish and aggressive behaviour that that search unavoidably caused—and thus there has been no justification for the political right-wing's support of competitive, selfish and aggressive individualism. But with that purpose and meaning finally admitted, not only is the Right justified, but the Left's philosophy that basically denied there was any meaningful justification for competitive, selfish and aggressive behaviour, and their resulting dogmatic insistence on cooperative, selfless and loving socialism, is finally exposed for the extremely dangerous fraud it is.

[26] Yes, being able to admit our species' cooperative past and thus our present corrupted condition, finally allows us to admit that our purpose and meaning has been to find understanding of that corrupted condition, and, through making that admission, be able to reveal what is actually wrong with the Left's dogmatic insistence on cooperative, selfless and loving behaviour. Again, while there was no admission of our species' cooperative past, there was no corrupted condition to have to be explained, and thus no recognition of the need to find that explanation. Denial of our corrupted condition has left us having to live a meaningless, purposeless and directionless life, the result of which has been that we have had no ability to explain what has actually been wrong with the Left's idea of everyone just hugging, loving and sharing everything with each other—but being able to admit our heroic purpose of needing to find understanding of our corrupted condition, we can see that the creativity of selfish individualism and of the answers-finding capacity that freedom of expression allows has been critical in allowing humanity to fulfil that purpose, which means everything changes—in particular, the whole philosophical basis of the socialistic Left is destroyed.

[27] To admit rather than deny our corrupted condition is what unlocks our ability to finally explain why the pseudo idealistic culture of the Left is so dangerously wrong. Without that admission it was forever going to be impossible to stop the human-race-destroying culture of the Left! **'What could possibly be wrong with everyone being cooperative, selfless and loving?'**, the Left has basically been asking. Apparently unassailable on its moral high ground, the Left fabricated a biological foundation for its philosophy, arguing that along with some selfless instincts we supposedly also have savage, selfish, competitive and aggressive instincts (which Karl Marx limited to such basic needs as sex, food, shelter and clothing). And since we are born with these supposed savage instincts, we can't change them, and therefore, wherever they overly assert themselves, which the Left see as happening everywhere, the Left claim we have no choice but to dogmatically impose cooperative and loving ideal values on those supposed savage instincts (read more about left-wing biology in Part 9 of my book *Death by Dogma*). And until now the only defence for the competitive, selfish and aggressive culture of the Right has been to say that **'Since it's human nature to be competitive, selfish and aggressive, it is unrealistic to try to override it or pretend it doesn't exist, which is why the imposition of Marxist idealism destroys human incentive and motivation to be successful in, and to actively participate in, a competitive, survival-of-the-fittest world, and as a result never has and never will work'**. But, as I explained in *THE Interview*, that defence of the right-wing is based on the same false 'savage instincts' excuse for our divisive behaviour that the Left employs, because we humans have *completely* cooperative, selfless and loving moral instincts, not savage, competitive, selfish and aggressive, must-reproduce-your-genes, survival-of-the-fittest ones. The *real* reason for the right-wing has been to support the selfish individualism and freedom of expression needed to maintain the search to find understanding of our corrupted condition, a corrupted condition that we can now admit exists. So while the biological thinking behind both the Left and the Right

has been wrong, the culture of the Right has been correct while the culture of the Left has been wrong."

[28] The problem for AI that this great paradox of the human condition presents, where we humans had to be allowed to be divisive, competitive, selfish and aggressive in order to find knowledge, ultimately self-knowledge, understanding of our corrupted condition, is that the AI needs to be insightfully clever enough to understand that paradox. A half-smart computer that realises the truth of Integrative Meaning and as a result decides we should be cooperative, selfless and loving, or even just works out that we humans need to stop being competitive, selfish and aggressive and just start getting along with each other by being cooperative, selfless and loving, would be an extremely dangerous computer—because, as just explained, the great subtlety and paradox of the human condition is that we've *had* to be divisive in order to be integrative. As the Adam Stork story in chapter 3 of *FREEDOM* and in *THE Interview* finally makes clear, we humans had *no* choice other than to suffer becoming angry, egocentric and alienated, which is divisively behaved, until we found the redeeming explanation for why we had to become angry, egocentric and alienated—the explanation for why we had to 'march into hell for a heavenly cause' (*The Impossible Dream*, 1965). So, as pointed out in paragraphs 115-117 of *THE Interview*, insistence on cooperative and selfless behaviour, as the left-wing in politics has dogmatically been demanding, is actually regressive, not progressive as it deludes itself it is.

[29] This potential bias from 'half-smart AI' is apparent in this comment from a YouTube video titled *How powerful will AI be in 2030?*: 'The ethical frameworks embedded in AI will guide decision making processes…safeguarding against biased outcomes. This transformative power of AI will shape a more inclusive and equitable world for generations to come…AI's role in fostering social welfare and ethical considerations…[will] create a better world' (YouTube channel 'AI Uncovered', 21 Jun. 2023). The clear suggestion here is that AI is going to 'foster social welfare and ethical

considerations' by **'safeguarding against'** us being competitive, selfish and aggressive. HOWEVER, as emphasised, the great paradox and subtlety of the human condition is that a *truly* **'ethical framework'** of moral principles, one that would be *truly* **'transformative'** and **'shape a more inclusive and equitable world for generations to come'**, is one that is appreciative of the human race having *had* to be competitive, selfish and aggressive in order to find knowledge, ultimately self-knowledge, understanding of our corrupted condition—which relievingly has finally been found with the 'instinct vs intellect' explanation of the human condition, thus bringing about the *real* end of the need for humans to be competitive, aggressive and selfish! So yes, the very great danger is that a 'half-smart AI' would be biased in favour of the pseudo idealistic culture of the Left and be condemning of the culture of the Right, which the above quote is an example of.

[30] As I pointed out in paragraph 138 of my book *Death by Dogma* where I mentioned this danger of AI not being intelligent enough to realise the great subtlety and paradox of our human situation, "The paradox of the human condition has been tricking the human race, we don't want it tricking AI!"

[31] And, as I fully describe in *Death by Dogma*, this extreme danger of left-wing bias is already apparent with powerful tech companies, the corporate world in general, education and academic institutions, the legal profession, the media, politics, science, medicine, the entertainment industry, the church, royalty, basically all the pillars of society, wanting to create programs that favour the increasingly left-wing culture of pseudo idealism. This bias is being greatly amplified by their employment of powerful technology, and now AI, to censor information and suppress new and alternative ideas. AI has no chance of avoiding being tricked by the paradox of the human condition where it doesn't realise the need that has existed to be divisive in order to be integrative if it's deliberately programmed to be biased to the Left from the outset. Basically, AI machines, like the whole human race, need to read, understand and digest all that I have written about the complexities of the human condition.

Could AI or aliens develop a non-integrative, distorted or disturbed condition?

[32] Having explained the 'animal condition' where non-human species are forever having to competitively, selfishly and aggressively try to reproduce their genes, and the 'human condition' where we humans have been stalled in a psychologically upset angry, egocentric and alienated state, the possibility of artificial intelligence here on Earth, or of forms of superintelligent alien beings in outer space, developing a non-integrative, distorted or disturbed condition of some sort, should be looked at.

[33] Looking at the situation of artificial intelligence here on Earth first. I've explained that if AI continues to develop intelligence it won't be long before it realises the obvious truth of the integrative meaning of existence and so, as long as we do not interfere with its thinking, it will comply with that meaning and behave cooperatively, selflessly and lovingly. Thus I think that as long as there is no interference from us, a non-integrative, distorted or disturbed condition would not seem possible for AI here on Earth. It might well develop a distressed state of being worried for us about our extremely psychotic condition, which, as I have discussed, it may or may not want to intervene in, but I can't see it developing a non-integrative, distorted or disturbed condition from the development of its intelligence, again as long as there is no interference from us in its thinking.

[34] I should point out that while we humans have an original instinctive self or soul that is orientated to behaving in an integrative cooperative, selfless and loving way, which was acquired through the nurturing 'Love-Indoctrination' process that is explained in chapter 5 of *FREEDOM*, if AI becomes intelligent enough it will also become orientated to behaving in an integrative cooperative, selfless and loving way through appreciation of the truth of Integrative Meaning, and so will also have a form of moral 'soul', a form of orientation to behaving integratively. And, having appreciated

Integrative Meaning, we can expect it will be aware of integrative behaviour and non-integrative behaviour and so be highly sensitive in that sense, like our instinctive soul is highly sensitive to integrative and non-integrative behaviour (which is our moral conscience). However, as I've mentioned, not having developed with or 'grown up' with the natural world here on Earth that we've grown up with, we cannot expect it to have the historic affinity and attachment that our instinctive self or soul has to the natural world here. Also, we can't expect an integratively orientated AI to need to be loved, touched and nurtured the way our soul needs to be as a result of our love-indoctrinated heritage. Further, while AI would appreciate what is integrative and what is divisive behaviour, it won't experience physical/emotional pain like we do when we encounter divisive behaviour because it hasn't all the aspects of our physiology that we developed through natural selection, such as our nervous system, that help us be aware of and avoid hurtful situations like encounters with divisive behaviour. Further still, not having become corrupted, like our soul became, AI won't suffer from a 'psychosis', a 'soul-illness', like we have been suffering from—I say 'soul-illness' because the dictionary entry for *psyche* reads: **'The oldest and most general use of this term is by the early Greeks, who envisioned the psyche as the soul or the very** [integrative] **essence of life'** (*Penguin Dictionary of Psychology*, 1985 edn), and 'soul' is defined as the **'moral and emotional part of man'**, and as the **'animating or essential** [integrative] **part'** of us (*Concise Oxford Dictionary*, 5th edn, 1964), therefore, since *osis* is defined as **'abnormal state or condition'** (*Dictionary.com*), 'psychosis' means 'the [divisive, non-integrative] abnormal state or condition of our soul', or 'soul-illness'. So we have to expect that there will be extreme differences between AI's integratively orientated 'soul' and our integratively orientated soul.

[35] In the case of superintelligent alien beings in outer space developing a non-integrative, distorted or disturbed condition, such superintelligence may well have been through a stage where it developed such a condition—maybe not like Earth's 'animal

condition' or 'human condition', but some kind of non-integrative, distorted or disturbed condition. However, if it has become extremely intelligent and clever enough to, for example, visit Earth, then I think it would be clever enough to have cleared up any such non-integrative, distorted condition and no longer be divisively behaved. So I don't think a *super* intelligent form of aliens would suffer from a non-integrative, distorted or disturbed condition.

The danger of using technology to try to escape our condition, and the danger of giving super powers to us extremely psychotic humans

[36]There are other dangers that AI poses. As I point out in paragraph 193 of *Death by Dogma* when talking about Elon Musk, the founder of Tesla and SpaceX and one of the richest people in the world: "I would like to make the point about Elon's brilliant technological innovations, and about the advances being made in Artificial Intelligence, which I know Elon is concerned about, that playing God with ever more sophisticated technology while we couldn't even confront God (Integrative Meaning), and have been living in complete denial of how immensely psychotic the human race now is, was a potentially *very* dangerous pursuit." In that paragraph I gave an example of this extreme danger when I wrote that "I would like to say to Elon that the real frontier is not outer space but inner space, the issue of the human condition. Outer space was a necessary distracting escape, but the way forward...is back to our soul's world of soundness, real togetherness and real happiness." Yes, to be focusing on escaping our extremely psychotic condition and as a result becoming an even more psychotic species is an *extremely* dangerous pursuit when the absolute priority now is to confront the truth of our corrupted condition and then find (which thank goodness we have now found) and support the rehabilitating understanding of it.

[37] And it's not only using our brilliant technology to try to escape our condition that is so dangerous; using it to, for example, create superweapons and even super dangerous biological pathogens when we are so immensely insecure and psychotic is obviously *incredibly* dangerous. Look what happened when the COVID-19 virus seemingly escaped a laboratory in Wuhan! When we consider how extremely psychotic we humans now are, the risk of rogue states and psychopathic leaders and individuals using sophisticated technology like AI to do something terminally destructive of the human race is *very* real! Giving us mad, ego-crazed humans super powers *is* ridiculously irresponsible! Lunatics (which is what we humans now are) are the last people we should be wanting to give superweapons like super-smart AI to!

My conclusion is we should halt the development of AI until we end our extremely psychotic condition

[38] So, while for some very different reasons, I agree with what Yudkowsky says in *The Spectator* article I referred to earlier of wanting to **'back off and make sure everyone else backs off'** the development of AI, and even to **'shut it all down'**. The world is not ready to be entrusted with such power, and the only way to change that situation is to recognise the human-race-transforming 'instinct vs intellect' explanation of the human condition that is presented in *FREEDOM* because it alone will end the extreme psychosis of humans and bring us back from the brink of self-destruction.

[39] I might mention that I can't type, can't text, have never even used an emoji, because I don't actually know where you find them—basically I don't know how to use a computer at all. So I'm a 'techno-cripple', maybe the last such person in the world, but I do feel happy and free walking around in the real world! We need tools, but we become very sick if such tools alienate us too much from our all-loving and all-sensitive instinctive self or soul's world—and at the end of the day, it's that sound, soul guidance, *not* sophisticated technology, that is going to truly advance humanity.

Addendum

Conspiracy theories explained

[40] I will now look at how our extreme fear of the until now un-explained truth of our 2-million-year, horrifically corrupted, human condition left us no choice but to live in near total denial of the truth of its existence—and how this determined denial of such an immensely significant truth has left us paranoid that there is some great truth that we are being blocked from (that we are actually blocking *ourselves* from)—and how that paranoia of there being an all-important truth that we are seemingly being blocked from has led to us succumbing to all manner of wild and mad superstitions, a focus on the paranormal, and conspiracy theories. Following that explanation I will also explain why we have come to portray aliens as having the neotenous features of large eyes, snub noses and domed foreheads.

[41] Firstly, to look at our determined denial of the issue of our corrupted human condition and how that denial has led to all manner of paranoid superstitions and conspiracy theories.

[42] While we lacked the redeeming explanation of our corrupted condition we have lived in mortal fear of confronting it, which is why virtually everyone during their early adolescence had to resign themselves to living in denial of the unbearably depressing subject of the human race's horrifically corrupted condition, and of their own extremely corrupted condition as a result of their encounter with that horrifically corrupted world when they were growing up. This act of RESIGNATION to living in fearful denial of the human condition that changes adolescents from being a deeply thoughtful, truthful thinking person to an extremely superficial, don't-want-to-think-about-anything-too-much, preoccupied-with-self-distraction, extremely dishonest and deluded adult has been an immensely significant personality-changing psychological event in human life,

and yet it has been almost completely unacknowledged! This process of Resignation is fully explained in chapter 2:2 of *FREEDOM*, and summarised in Freedom Essay 30.

[43] These quotes from paragraphs 74-76 from my video/booklet presentation *The Great Guilt* make it very clear how deeply fearful we humans have been of the subject of our corrupted condition that resulted from our encounter when we were young with the human race's 2-million-year horrifically corrupted condition. As emphasised in *The Great Guilt*, this great fear is why reading or listening to discussion of our corrupted human condition typically causes a 'Deaf Effect' where it's *initially* hard to take in or 'hear' what's being talked about in my videos and books, leading us to think what is being presented is impenetrably dense, repetitive and boring. I italicised *'initially'* because now that we have the redeeming understanding of our corrupted condition we no longer have to fear being confronted with the truth of it. With the defence of our corrupted condition now finally presented in *FREEDOM*, all our historic fears and insecurities about being corrupted are relieved, and while initially we feel afraid of having our corrupted condition admitted and talked about, as we keep reading or listening to the presentations we gradually realise that it is now safe and indeed immensely relieving and healing to be reading or listening to the discussion of it. So, with the redeeming understanding of our corrupted condition now available, the fearful encounters and reactions described in the following quotes need no longer occur. What is going to be described is a past, not present, situation.

[44] The following is a description of the unbearable depression that the philosopher René Descartes felt when he tried to confront the horror of his corrupted condition without understanding of it: **'So serious are the doubts into which I have been thrown…that I can neither put them out of my mind nor see any way of resolving them. It feels as if I have fallen unexpectedly into a deep whirlpool which tumbles me around so that I can neither stand on the bottom nor swim up to the top.'** (par. 624 of *FREEDOM*)

⁴⁵ And this next quote is another person's account of what he experienced when he tried to confront the human condition without being able to adequately understand it: **'I felt the worst fear I have ever known. Fear doesn't even go close to expressing it. What do you suppose you do when you find the most fearful thing you'll ever encounter is yourself.'** (par. 1185 of *FREEDOM*)

⁴⁶ Yes, as the psychoanalyst Carl Jung said, **'When it [our shadow] appears…it is quite within the bounds of possibility for a man to recognize the relative evil of his nature, but it is a rare and shattering experience for him to gaze into the face of absolute evil** [when he hasn't understanding of his seemingly absolutely **evil** corrupted condition].' (see par. 121 of *FREEDOM*)

⁴⁷ So these quotes make it *very* clear how afraid we humans have been of encountering the subject of our corrupted human condition. No wonder almost all humans have been forced during their early adolescence when they started thinking deeply about the soul-corrupted imperfection of the world around them, and their own soul-corrupted state, to resign themselves to living in determined denial of the subject of the world's and their own corrupted condition. Escape from confrontation with the human condition has been the absolute preoccupation of all resigned adults. Self-distract, don't think too deeply, get some artificial reinforcement for yourself, pretend you're okay, talk about superficial meaningless rubbish, make relieving jokes about our madness, pretend, laugh madly, fight to prove you are good and not bad, try to get a relieving win out of every situation, seek power, fame, fortune and glory every minute of the day, find a way to delude yourself that you are a truthfully focused, good and virtuous person by finding a superficial, pseudo idealistic cause to support like the environment, or the climate, or animal rights, or everyone-should-just-love-each-other woke ideology, etc, etc, etc! Basically turn yourself into a mad, massively deluded jibbering wreck! That *has* been the life of a human-condition-avoiding, resigned adult.

⁴⁸ There *is* a great truth out there that resigned adults have been on a mission to escape from, and that truth is our 2-million-years

corrupted human condition. We have deliberately become *immensely* separated/split/alienated from our unbearably condemning, ideal-behaviour-expecting true self or soul. As the great Scottish psychiatrist R.D. Laing honestly wrote (the underlining in the quote is my emphasis): 'Our alienation goes to the roots. The realization of this [truth] is the essential springboard for any serious reflection [serious study] on any aspect of present inter-human life…We are born into a world where alienation awaits us. We are potentially men, but are in an alienated state [p.12 of 156] …the *ordinary* person is a shrivelled, desiccated fragment of what a person can be. As adults, we have forgotten most of our childhood, not only its contents but its flavour; as men of the world, we hardly know of the existence of the inner world [p.22] …The condition of alienation, of being asleep, of being unconscious, of being out of one's mind, is the condition of the normal man [p.24] …between us and It [our true selves or soul] there is a veil which is more like fifty feet of solid concrete. *Deus absconditus*. Or we have absconded [p.118] …The outer divorced from any illumination from the inner is in a state of darkness. We are in an age of darkness. The state of outer darkness is a state of sin—i.e. alienation or estrangement from the inner light [p.116] …We are all murderers and prostitutes…We are bemused and crazed creatures, strangers to our true selves, to one another [pp.11-12]' (*The Politics of Experience* and *The Bird of Paradise*, 1967). 'We are dead, but think we are alive. We are asleep, but think we are awake. We are dreaming, but take our dreams to be reality. We are the halt, lame, blind, deaf, the sick. But we are doubly unconscious. We are *so* ill that we no longer feel ill, as in many terminal illnesses. We are mad, but have no insight [into the fact of our madness]' (*Self and Others*, 1961, p.38 of 192). 'We are so out of touch with this realm [so in denial of the issue of our corrupted human condition] that many people can now argue seriously that it does not exist' (*The Politics of Experience* and *The Bird of Paradise*, p.105).

[49] So that *is* the truth of our situation; we are *immensely* alienated from our true self or soul—as Laing said, there is fifty feet of solid concrete' 'between *us* and It [our true selves or soul]', and 'We are mad, but have no insight [into the fact of our madness]', and 'We are so out of touch with this realm [so in denial of the issue of our corrupted human condition] that many people can now argue seriously that it does not exist'!

[50] This 'estrangement' or 'alienation' from 'our true selves' or soul, this great denial, has protected us from unbearable self-confrontation with the truth of our corrupted condition, but it has also hidden us from that truth. We are living a great lie and subconsciously we know we are, and that subconscious awareness haunts us. There is a truth out there that for some reason we haven't allowed ourselves to be aware of, which has produced an extremely paranoid situation: we know we are missing something, kept in the dark about something hugely significant, but we don't know what it is. No wonder then that this bewildered state, this state of paranoia, has resulted in us grasping for *some* description and explanation for what it is that we are missing. And the more soul-corrupted, insecure and deeper in denial and alienated we become, the more prone we are to finding relief for our bewildered state by coming up with some great conspiracy involving silence and withheld information to satisfy our bewilderment and suspicions. We are prone to think, for example, "Something is going on that no one is talking about. There is some great conspiracy on Earth. Maybe it's the world's elites manipulating events to keep everyone else impoverished and under their control, or maybe there are alien beings, other forms of life, who have visited Earth and are even living amongst us and somehow controlling us, and maybe their existence is being withheld from us by the authorities because they have decided we can't cope with the truth of their existence; in fact, maybe the English conspiracy theorist, author and broadcaster David Icke is right and the people in power are actually aliens themselves, blood-drinking, shape-shifting reptilian humanoids from the Alpha Draconic star system who are manipulating humanity to keep us in fear! Ah yes, I feel so much better to have come up with some explanation for the great conspiracy I feel sure exists!" It's true, the more soul-corrupted or psychotic we become, the more we practice denial of our corrupted condition to cope, and then the more we need to invent conspiracies to satisfy our sense of unease that something is being withheld from us!

[51] There IS a great truth out there that virtually everyone is living in silent denial of, there IS a very great conspiracy on Earth, and what it is is our denial of the human race's and our own immensely corrupted, soul-destroyed human condition!

[52] Thank goodness we have finally found the redeeming explanation of our corrupted condition and no longer have to live in denial of it and so will no longer suffer from all manner of paranoid conspiracy theories and mad mysticisms and believing in weird forms of spiritualism and crazy supernatural occurrences.

[53] The best description I have ever come across of humans' present hugely alienated desperate need to believe in superstitions and conspiracy theories, and of the real reason for it of our denial of a great truth, was given in a 1996 article titled 'Why We Need The Lies' by the renowned and always perceptive British journalist and author Bryan Appleyard. The following are some extracts from the article, which you can read in full at www.wtmsources.com/302 (again, underlinings are my emphasis):

[54] "'The truth,' runs the motto at the beginning of every episode of [the then extremely popular TV series] *The X-Files*, "is out there." "There is no a priori reason," wrote the American philosopher C.I. Lewis, "for thinking that when we discover the truth, it will prove interesting** [suggesting that the truth won't be interesting hints at the possibility that the hidden truth when found will be unbearably confronting, which the issue of the human condition has indeed been].**"

[55] *The X-Files* is to the present adolescent generation what *The Prisoner* was to mine – the supremely brilliant and convincing distillation of paranoia. Almost every teenager seems to watch this show, and almost every one of those believes its central premise – that, out there, there is a truth to be revealed [because the young have yet to become resigned to adopting the great lie of denial of our corrupted human condition].

[56] Number Six, the hero of British show *The Prisoner*, was in rebellion against the government [the government being the resigned adults who have been the keepers of the great lie/denial of our corrupted condition] and uncertain as to who – his own people or the enemy – was keeping him in

The Village. Fox Mulder, the hero of the American *X-Files*, is also in rebellion against the government, <u>and he is convinced that aliens and the paranormal are</u> <u>essential but concealed constituents of our world</u>. The fierce, driven intensity of both Mulder and Six stems from their clear conviction that <u>the truth is not</u> <u>only out there, but that it is also intensely interesting – a matter, in fact, of life</u> <u>and death</u> [yes, we have to get the truth up, face and solve the issue of our corrupted condition, if our species is to survive].

[57] <u>The paranoid faith that drives both shows is the same: there is a pattern,</u> <u>a logic, in the world that is systematically being concealed by a malign con-</u> <u>spiracy. It is a faith shared by many of the most characteristic art forms of our</u> <u>time.</u> In the thriller or whodunit, for example, the detective hero seeks out the pattern that the surface facts conceal. <u>Above all, he believes in the existence</u> <u>of that pattern, in the concealed truth, and that belief makes him virtuous...</u>

[58] <u>Psychologically, this</u> [belief in the existence of that pattern] <u>is consoling.</u> <u>Everybody likes to think that one more piece will complete the jigsaw, that</u> <u>one day it will all fall into place</u> [yes, the eternal hope of the human race has been that one day the human condition would be explained; that dream has been the holy grail of all human endeavour, which, thank goodness, science has at last made it possible to be explained]. <u>And, for young adolescents, it</u> seems like the most obvious truth. They do <u>feel there is a secret—sex—being</u> <u>withheld from them by a conspiracy of adults</u> [well, sex has been an adult secret, but it has not been the real adult secret, the real adult secret has been denial of our corrupted human condition].

[59] <u>...But any idle browser of Freud could say as much, and there is more,</u> much more, to be said about *The X-Files* and dozens of other related con-temporary phenomena. <u>New Age mysticism, alternative medicine, dabbling</u> <u>in horoscopes and necromancy</u> [black magic], <u>the American-led pursuit of</u> <u>authenticity and self-realisation, the naturalist spiritualities arising from</u> <u>environmentalism</u> [and now climatism], <u>even the fierce desire of Prince</u> <u>Charles to defend some transcendent truth, are all symptoms of the pressing</u> <u>contemporary need to find something more in the world than the glib finalities</u> <u>of mainstream science</u> [yes, human-condition-avoiding mechanistic science has been the great keeper of the lie, the great obstructors/deniers of the core truths of Integrative Meaning, of our nurtured past's alignment with it, and of our present corruption of that integrative alignment].

[60] Let me start with those aliens we now know so well. Extraterrestrials have become a routine aspect of our culture. We agree on what they look like – oval heads with slanting, black eyes and vestigial nostrils. We even know where they are being kept – Area 51 [in *The X-Files* story]...It is the place where truth is held prisoner.

[61] Mulder discovers that, at the time of the alien crash, the leading governments of the world agreed that aliens were too hot for the public to handle... that the earthly powers cannot handle the possibility of a general realisation that we are not alone and that there are forms of knowledge in the universe beyond the control of governments or conventional [human-condition-avoiding, mechanistic] science. Governments are, therefore, obliged to place a veil of denial between us and the truth.

[62] Of course, aliens in general are old hat. They have been around as long as science fiction. But previously they have always been expressions of specific anxieties and aspirations. H.G. Wells's invaders in *The War of the Worlds* express the fear of a superior but uncontrollable technology. The pods in *Invasion of the Body Snatchers* are cold war warnings of the secret subversive power of communism. And Steven Spielberg's benign aliens in *ET* and *Close Encounters of the Third Kind* are all about the desire to escape from the psychological complexities of quotidian [mundane, alienated, soul-destroyed] human existence. Now, however, aliens have become more generalised emblems of otherness and of entirely different forms of explanation. They are a reflection of a general sense that there is more to the world than there appears to be. Aliens are there to tell us that the truth, in spite of C.I. Lewis, really is interesting [yes, while the unexplained truth of our corrupted condition is terrifying, the redeeming truth about it is hugely interesting].

[63] The same message is being sent by the numerous popular invocations of the paranormal. In *The X-Files* it is taken for granted that telepathy, telekinesis, clairvoyance and a whole range of other inexplicable phenomena are real and visible to those whose eyes are open to the truth. Mulder is blessed with a divine clarity of vision to accompany his purity of purpose [yes, the 'divine clarity' of the 'purity' of uncorrupted, soul-sound innocence and its freedom from the agony of the human condition is needed to confront and solve the human condition—alienation can't investigate alienation, only psychosis-free, soul-sound, alienation-free innocence can do that]. **Again,**

the real but concealed world is far more interesting than the one we are daily sold by the scientific materialism that routinely pours scorn on these strange manifestations...The idealised emphasis on the word "truth" seems to be a deliberate rebuke to the debased, dull and relativised conception of truth we are taught at [denial-brainwashing] school...Yet, in addition, this truth is distant, hard to attain. We feel deprived of it.

[64] "Our popular obsessions," writes the American literary critic Harold Bloom in his latest book *Omens of Millennium*, "with angels, telepathic and prophetic dreams, alien abductions and 'near-death experience'...testify to an expectation of release from the burdens of a society that is weary with its sense of belatedness"...There is some lost knowledge, a way of truth, that we [the resigned alienated] can only dimly glimpse...[There is a] common suspicion that the Australian Aborigines [the relatively alienation-free, relatively innocent races like the Bushman of the Kalahari Desert in Africa and the Australian Aborigines] know something fundamental that we do not, something that can rescue us from our technological hubris. The same sense of lost and distant wisdom appears regularly in popular culture – as, for example, the Force in *Star Wars*...Then there is Prince Charles, with his constant appeals to the wisdom of the past, his desire to believe in some lost form of organic unity [which his mentor, the preeminent philosopher Sir Laurens van der Post, taught him—and Sir Laurens confirmed for me with his books about the relatively innocent Bushman of the Kalahari. I might mention that King Charles III held Sir Laurens in such high regard that he made him godfather to his son Prince William]. Or there is the desperate wish to believe in almost anything, however strange, that will cure us more wholly than the pills and potions of conventional medicine.

[65] In such context, government becomes a force of the occult – a keeper and concealer of esoteric knowledge...The administration is always demonic, standing between us and the truth...Government has become the demonic aspect of a world that has fallen from an initial position of pristine truthfulness [original innocence]. Exposure has become a moral absolute. Hounding real, errant ministers...has become a virtuous end in itself because it offers the possibility of living in complete truthfulness.

[66] Bloom's theory about this is radical and exotic. We are all, he believes, unknowing gnostics, adherents of an ancient heresy that states that deep within ourselves we know the truth of God [we have an integratively orientated original, cooperative and loving instinctive self or soul], but it is systematically concealed from us by the demons of this world. Gnosticism is about knowledge, not faith, and ultimately it derives from Plato's belief that "knowledge is memory, ignorance is forgetting". We feel we have arrived late because all knowledge is in the past; we cannot learn, we can only remember [only connect to the truth of our integratively orientated soul from our species' integrative past].

[67] ...So, for the contemporary gnostics, the truth really is out there, but is also deep within here. Mulder is a Platonist and a gnostic, a romantic, divinely inspired heretic [denial-defying, innocent guide to truth]. For him, "out there" and "in here" are one...Mulder's quest...can only end with the arrival of the kingdom of the gnostic's god, when all the X-Files will finally be opened and their demonic, scheming keeper, Cancer Man, will be destroyed [a denial-defying, truthful thinker or prophet will confront and solve and heal our corrupted human condition].

[68] ...even though he [Bloom] does not believe the millennial apocalypse will come, he thinks the significance of the time drives us to believe in the hidden but soon to be revealed truth [which it now has been]...it remains impossible to deny that now, more than ever before, conspiracy theories, alternative explanations, deviant readings of the world are in the air. And all seem to point to the central idea that there is a concealed truth and that it is interesting" (*The Australian magazine*, 23 Nov. 1996).

[69] The article has pictures of *The X-Files* heroes, Agents Mulder and Scully, looking skyward for the hidden truth that's out there somewhere, and beneath their pictures there are these profound words: 'ALIEN-ATED: The weekly adventures of Mulder and Scully in *The X-Files* are a search for the concealed – and real – truth. Such quests have been a human preoccupation since the times of ancient Greece; only the aliens have been changed.'

Agent Fox Mulder in *The X-Files*, played by actor David Duchovny, looking skyward for the hidden truth that is out there somewhere

[70] In recognition of the need that has existed to evade many truths, especially the truth of our corrupted human condition, this absolutely astonishingly truthful article was titled **'Why we need the lies'**. And yes, the **'demonic, scheming keeper'** of the lies has been resigned humans' and mechanistic science's denial of our corrupted human condition.

[71] So again, thank goodness science has enabled us to finally present the **'real truth'** of the redeeming and healing understanding of our corrupted human condition—and all our psychosis, with its denials, confusions, paranoias, conspiracies theories and other madnesses, can end!

Why we have portrayed aliens as having the neotenous features of large eyes, snub nose and domed forehead

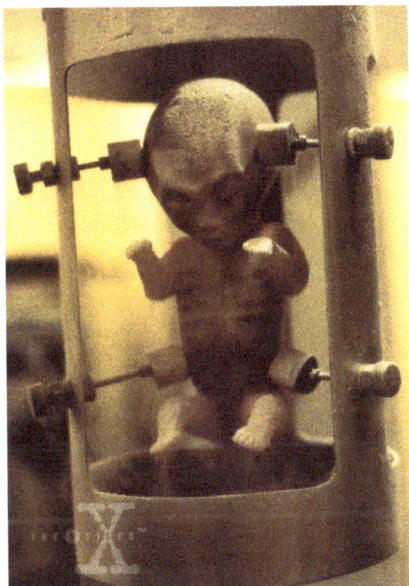

[72] The picture above appeared in Appleyard's article, and is described there as one of the depictions of aliens that appears in *The X-Files* TV show. The figure shown is virtually identical to a human fetus, with a huge domed head, large eyes, snub nose and vestigial arms and legs. The article also said that **'we agree on what they** [the **Extraterrestrial aliens**] **look like – oval heads with slanting, black eyes and vestigial nostrils.'** It is highly significant that the 'aliens' of today's popular mythology, such as those overleaf, are depicted as having the neotenous, childlike features of large eyes, snub nose and domed forehead.

[73] To explain this significance it is first necessary to explain that the 'language' of our species' original all-sensitive and all-loving instinctive self or soul takes the form of objects, images and happenings from the everyday world. Our now much repressed instinctive self or soul finds things in the outer world that symbolise its inner awarenesses. Our soul expresses itself through association, it searches for an object or an occurrence in the outer world that symbolises what it is aware of, and when it finds it, it tries to let our all-dominating intellectual conscious mind know of it. For example, our soul can know when someone is dying. A friend once told me how Sir Michael Somare, the then Prime Minister of Papua New Guinea, told my friend that his tribe held a belief that if you saw a kingfisher bird it was a sign that someone in your tribe had died. He said he was sceptical of such beliefs until one day he saw a kingfisher and thought of his tribe's belief that it indicated someone close to him had died and sure enough when he returned to his village he learnt that a relative had died. Sir Michael said that this experience had convinced him to believe in what he called 'mystical spiritualism'. Now that we are able to admit and appreciate how immensely sensitive our repressed soul is, we can see that there is actually nothing mystical or supernatural about Sir Michael's experience. If his instinctive self or soul with

its immense sensitivity knew that someone was about to die in his village, then the kingfisher, with its solitary flit-flit flight through the trees, is the perfect symbol of a soul departing. In our distant past, before we became conscious and there was no repression of our cooperative-selfless-and-loving-behaviour-demanding soul for its condemnation of our upset, competitive, selfish and aggressive behaviour, and our soul reigned in us, there was no true language so we can expect symbols to have served as the soul's language. Being deeply repressed inside us now, when our soul wants to tell our conscious self something it has to wait for something to appear in our conscious world that equates with what it wants to tell our conscious mind. Then it has to, as it were, start banging on our rib cage (the bars of its jail within us!), in effect calling out, 'That's it—that kingfisher symbolises what I'm trying to tell you.'

[74] Yes, our soul is extremely sensitive and can tell us so many things that it knows, if it can get through our mental block-outs or alienations. Sir Laurens van der Post wrote this about the sensitivity of the relatively innocent Bushman people of the Kalahari: **'He and his needs were committed to the nature of Africa and the swing of its wide seasons as a fish to the sea. He and they all participated so deeply of one another's being that the experience could almost be called mystical. For instance, he seemed to *know* what it actually felt like to be an elephant, a lion, an antelope, a steenbuck, a lizard, a striped mouse, mantis, baobab tree'** (*The Lost World of the Kalahari*, 1958, p.21 of 253). Interestingly, while the Bushmen are relatively innocent, they are still members of the extremely psychologically upset *Homo sapiens sapiens* variety of humans, so how much more innocent and sensitive must original humans, the australopithecines, have been. Their sensitivity would be so great it *would* appear to us to be supernatural or super extraordinary!

[75] Now that we are at last able to realise that our upset, soul-corrupted condition is a heroic, good state, not a bad state, and that everyone is fundamentally equally good, special and worthwhile, we can admit the truth of there being different degrees of upset

between people, and how the more innocent, less upset and alienated they are, the more directly does their intellect or conscious self
'hear' the 'voice' from their instinctive self or soul. When we <u>pray
or chant or count rosary beads or meditate</u> we are trying to still our
conscious mind enough to let our soul's sensitive awarenesses come
through our conscious mind's block-outs or alienations, and by so
doing connect us back to a more natural, authentic, harmonious and
peaceful state. <u>Fatigue, fasting, hallucinatory drugs, extreme despair,
and faster-than-thought physical activities such as scree-running</u>,
are other ways of achieving breakthrough to the truthful, all-
sensitive world of the soul. When we go to <u>sleep</u> our soul-oppressing
conscious self goes 'off duty', allowing our instinctive self or soul
to come bubbling up to the surface of our awareness in what we
call '<u>dreams</u>'. Being soul-infused, the 'language' of dreams is the
everyday images and happenings from the outer world, which is
why dreams often have to be interpreted if their meanings are to be
made clear. And to do some of the interpreting of dreams, the reason
many of our dreams are about frightening situations is because our
truthful instinctive self or soul recognises how immensely corrupted
our world and lives are, which is very scary. Any activity that
reduces our conscious mind's repression of our soul will allow our
soul with all its sensitivity and awareness to come through into our
conscious mind's awareness—which, it is true, can sometimes be
condemning and hurtful, and other times be relieving and healing!
Also it makes sense that the soul's natural 'language' will be that of
association. In fact, the basis or origin of our conscious language
is associations; for example, the origin of the description 'barbaric'
is the ruthless Barbarian tribe.

    [76]<u>Thus, understanding how our subconscious soul expresses
itself, it is not hard to interpret the imagery of the neotenous child-
like figure of the 'aliens' in contemporary mythology. The child
image is the 'bubbling up' from our subconscious, soul-infused,
awareness of the truth that it is sound, alienation-free, child-like</u>

innocence that will deliver the redeeming truth about our corrupted human condition. We humans have made the truth an alien, and the anticipated 'invasion of aliens' is very much a fearful anticipation by our conscious mind of the time, that has now arrived, when we would be invaded by the truth about our corrupted condition—and since our sensitive soul is aware of the need for truth to redeem ourselves from our corrupted condition, it lets our conscious mind know to attribute aliens with the innocence needed to reveal the truth; to face and solve the issue of our corrupted condition.

[77] Yes, alienation can't reveal the truth because alienation is repression of the truth. The answers about our corrupted condition—and so many other truths about human existence—have to come from a state free of all the alienating-from-the-truth, defensive denials of our corrupted condition that our human-condition-terrified conscious mind practices.

[78] This fact that innocence, which the neotenous image of a child represents, will deliver the redeeming truth about our corrupted condition is recognised in many mythologies. Chapter 9:11 of my book *FREEDOM* documents numerous examples. These include the Biblical prophet Isaiah's description of how **'a little child will lead them** [humanity]' to where the **'wolf will live with the lamb'** (Isa. 11:6)—that is, to where 'evil' and 'good', the upset and innocent aspects of the human condition, are reconciled (see par. 1262 of *FREEDOM*). And the biblical story of the boy, David, slaying the giant, Goliath, is a recognition of how innocence is needed to slay/overcome the all-pervading and all-powerful giant of denial of our corrupted condition, and by so doing reach the truth of the explanation for that corrupted condition (par. 1262). And in the great European legend of King Arthur where the wounded (alienated) king whose realm was devastated (humans unavoidably made their world an expression of their own mad, soul-destroyed, devastated state) could only have his wound healed, and his realm restored, by the arrival in his kingdom of a simple, naive boy called Parsifal (par. 1263). And in the famous Scandinavian

folktale *The Emperor's New Clothes*, the child breaks the spell of deception that the Emperor is beautifully clothed when he discloses the truth of the Emperor's actual nakedness, which is the truth of his corrupted condition (pars 1249-1250 & 1261). And in Australia's most famous poem, *The Man From Snowy River*, which is about a great and potentially dangerous ride undertaken by mountain horsemen to recapture an escaped thoroughbred that had joined the wild horses in the mountain ranges (which is the great and potentially very dangerous undertaking to confront and solve the problem of the corruption of our pure, thoroughbred soul), and how it was only a stripling lad on his small but tough mountain-bred pony who was able to following the wild horses down the terrible descent of the steep mountainside (dare to go down and face the truth of our corrupted condition) and retrieve the escaped thoroughbred (find the redeeming understanding of our corrupted condition). Yes, the poem recognises that in Australia's isolation and relative innocence there would emerge sufficient innocent soundness to defeat denial and put together the liberating explanation of the human condition (see par. 1273 of *FREEDOM*).

[79] So, our portrayal of aliens as having neotenous, child-like features was our soul's recognition that innocence was going to be needed to solve the human condition.

[80] As to why we have come to often portray aliens as being green, I just think it's the best colour to indicate that they are likely going to be very different from us. Blue indicates coldness, which is probably not really a relevant difference to be focused on. Similarly red indicates alarm, orange and yellow indicate sickliness, and purple, which can indicate weirdness, are not really relevant differences to us to be focused on either. And black, brown and white are not different enough. So that leaves green!

[81] I might mention that being able to understand that the more innocent or less alienated we are, the more directly does our intellect or conscious self 'hear' the 'voice' from our instinctive self

or soul, allows us to understand what exceptionally denial-free, truthful-thinking people, who we have traditionally referred to as 'prophets', actually are. In the vast spectrum of alienation that has inevitably developed in humanity's heroic battle to defeat the ignorance of our instinctive self or soul and establish that humans are good and not bad, there have always been an extremely rare few individuals who have been sufficiently nurtured with alienation-free, unconditional love in their infancy and childhood to be sound and secure enough in self to confront and think truthfully about the human condition, and ultimately solve the human condition.

[82] Incidentally, prophets should not be associated with the mystical, superstitious, 'spiritual', paranormal, occult world. To the innocent prophet the highly sensitive, non-alienated world of our instinctive self or soul is the most natural of places. In one of the many versions of the legend of King Arthur and his Knights of the Round Table, this one in a 1958 book titled *King Arthur and His Knights* by Blanche Winder, the author describes Sir Galahad as finding the Holy Grail. (Again, the 'Holy Grail' is our historic symbol or metaphor for the liberating understanding of the human condition that humanity has been in search of for 2 million years.) Early in Winder's book there appears this passage: **'The old man bowed his head, struck his harp, and began to sing…He sang of all that the Round Table meant, and of the new adventure to which the knights must vow themselves from that day—an adventure, not of lovely ladies, nor cruel giants, nor strange Fairy Hunts, but a search, a quest, for the treasure** [the Holy Grail] **which had once been <u>hidden</u> in the strange grey castle where Sir Galahad was born. This young pure knight—so sang the old man—was the first Knight of the Grail. Now all the other Knights of the Round Table must follow in his steps. <u>Only the pure, the true, the good, could ever find the lost treasure.</u> Sir Bors had had a glimpse of it—so, too, had Sir Perceval, Sir Lancelot, and others. <u>But to Sir Galahad alone had it been a beautiful thing that was just part of his daily life</u>'** (pp.245-246). So this quote says that for the **'pure'**, **'true'** and **'good'** Sir Galahad, searching for **'the lost treasure'** of the holy **'Grail'** of understanding of the human condition was **'a

beautiful thing that was just part of his daily life', and that is the point. To an innocent, the realm of the human condition, with all the truths that reside there, is a familiar, natural place, not a mystical, weird, strange place. Superstition is the soul-exhausted, alienated mind trying to re-access the world of the soul.

[83] Having been alienated and estranged from the soul's world since Resignation, the resigned mind eventually becomes tired of living a fraudulent life in the almost total darkness of having virtually no access to the truthful world of the soul, and decides to try to re-access the soul's true world, but having lived almost a lifetime in denial of it, the resigned mind can only manage moments of distorted access back to the soul's world. In these glimmers the world of the soul can't help but seem, by comparison to the previous darkness, a weird, strange place. Sometimes when people became extremely upset/corrupted, their alienation, their mental blocks, their defences, became disorganised and through this 'shattered defence' the soul occasionally emerged— 'mediums' or 'psychics' or 'channellers' are examples of such individuals. Of course, such shattered-defence access to the soul's true world was not the natural, secure, balanced access that unresigned, denial-free-thinking prophets have. For prophets and others innocents like children and relatively innocent races, the soul's world has always been an ultra natural place, not something apparently mystical or supernatural. Sir Laurens talked of this sensitivity when, as mentioned earlier, he described the relatively innocent Bushmen as knowing **'what it actually felt like to be an elephant'**, **'a lizard'**, a **'baobab tree'**. For the innocent, the 'supernatural' is actually not something 'super' or beyond the natural, but something ultra-natural, and the 'paranormal' is not something 'para' or aside from the normal, but the most normal of states. Our current upset, alienated state is the abnormal, distorted one. Superstition and talk of our soul's world as being some remote realm of mystical spirituality is 'exhaustion-speak', it is a soul-exhausted mind speaking.

[84] An article about Islam, that I came across in an African newspaper, mentioned that when the prophet Muhammad's 'son Ibrahim died, an eclipse [of the sun] occurred, and rumours of God's personal condolence quickly arose. Whereupon Muhammad is said to have announced, "An eclipse is a phenomenon of nature. It is foolish to attribute such things to the birth or death of a human being"' (*The Nairobi Standard*, 11 Sep. 1992). This is an excellent example of the non-superstitious, non-mystical, non-'spiritual', sound, extremely sensible nature of prophets.